Habitats and Change

Reader

Copyright © 2019 Core Knowledge Foundation
www.coreknowledge.org

All Rights Reserved.

Core Knowledge®, Core Knowledge Curriculum Series™, Core Knowledge Science™, and CKSci™ are trademarks of the Core Knowledge Foundation.

Trademarks and trade names are shown in this book strictly for illustrative and educational purposes and are the property of their respective owners. References herein should not be regarded as affecting the validity of said trademarks and trade names.

Printed in Canada

ISBN: 978-1-68380-510-6

Habitats and Change

Table of Contents

Chapter 1	**Habitats**	1
Chapter 2	**Adaptations in Living Things**	7
Chapter 3	**Living Things in Groups**	13
Chapter 4	**Nature and Changing Ecosystems**	19
Chapter 5	**Humans Cause Ecosystem Change**	25
Chapter 6	**Living Things Respond to Change**	31
Chapter 7	**Ecosystem Problems and Solutions**	35
Chapter 8	**Fossils and How They Form**	41
Chapter 9	**Fossil Clues About Changing Habitats**	47
Chapter 10	**Habitat Changes and Extinction**	53
Glossary		59

Habitats

Chapter 1

Living things need food, water, and shelter. All living things must live in places where they can get what they need. A **habitat** is the natural place where an organism lives. A habitat provides an organism with what it needs to **survive**, or stay alive.

There are many parts that make up a habitat. For example, land and water are parts of a habitat. Land can provide organisms with things like food and shelter.

Weather is part of a habitat, too. Temperatures make places livable or not livable. If places get too hot or too cold, organisms might not be able to survive there.

Big Question
What do all habitats have in common?

Vocabulary
habitat, n. the natural place where an organism lives

survive, v. to stay alive

A coral reef in the ocean is a clownfish's habitat. The fish can get food there. It has a place where it is protected from being eaten by other animals.

Very Cold Habitats

Some habitats are very cold. The land in cold habitats is sometimes rocky, with few trees or plants. It can be snowy and windy there. Mountains and tundras are two kinds of cold habitats.

The arctic hare lives in a cold habitat.

An arctic hare is an animal that lives in a very cold habitat. It can survive there because it has fur that keeps it warm. The color of its fur helps it blend in with its surroundings. It eats the kind of plants that live in cold habitats.

Word to Know

A *tundra* is a cold habitat that is rocky and flat. The soil is frozen, so only a few kinds of plants can survive there. The weather in a tundra is harsh.

Oceans can be very cold, too. These cold oceans may have blocks of ice floating in them. Blue whales, narwhals, and humpback whales are a few of the animals that live in cold ocean habitats.

Some places in oceans are so cold that ice floats there.

Very Dry Habitats

Some habitats are very dry. Dry habitats can be hot or cold. They can be flat or mountainous. A desert is one kind of dry habitat. A tundra is a dry habitat, too. Deserts and tundras get much less rain than other kinds of habitats.

Dry habitats usually have few plants. Most plants cannot survive in the harsh conditions. Many members of the cactus family live in deserts. They do not need much water to survive.

The animal in the picture is a kit fox. The kit fox can hear its prey moving around under the ground. Its large ears are good for more than hearing. They help it release heat from its body in its hot, dry habitat.

Desert plants can survive with little water.

The kit fox lives in a hot, dry habitat.

Word to Know

A *desert* is a dry habitat. It can be hot or cold depending on where it is. A desert's soil is usually sandy and rocky. Some deserts appear to be nothing but sand.

Forest Habitats

Forest habitats have many trees. In some forest habitats, the trees have broad leaves. In others, the trees have leaves that are needle-shaped. Some forest habitats have both kinds of trees. Many other organisms live in and around the trees of a forest.

This forest is mostly evergreen trees.

Forests can be warm or cool. Most get plenty of rain and snow throughout the year. Rain forests are wet all year long. Another type of forest is the temperate forest. Temperate forests are warm or cool depending on the season. Temperate forests get both rain and snow but less rain than rain forests get.

Forests provide food, water, and shelter for many organisms. Deer graze on the shrubs under tall trees. Birds and other animals take shelter in the trees. Mushrooms grow in the damp soil.

A raccoon is one of many animals that live in a forest habitat.

Water Habitats

When you think of a water habitat, you might picture an ocean. Or you might think of a lake, pond, river, or stream. All of these are water habitats.

Almost all of Earth's water is found in oceans. Ocean water is salty. It can be warm or cold. Animals such as seals, urchins, and coral live in ocean habitats. Kelp and other kinds of algae live in ocean habitats.

Kelp grows underwater in oceans.

Ponds and lakes have land on all sides. The water in ponds and lakes is mostly still. Rivers and streams have water that is flowing. Unlike oceans, these habitats are fresh water, not salty. Fish, turtles, and ducks are some animals that live in freshwater habitats.

These turtles live in a pond habitat. They come out of the water to warm themselves in the sunlight.

Habitats and Ecosystems

Habitats may have different land and water features and different kinds of weather. However, all habitats have one thing in common. They meet the needs of the organisms that live there. If organisms do not get what they need, they will not survive.

Organisms in a habitat interact with each other. They also interact with the land and water around them. Together, all of these things form an ecosystem. An ecosystem is made up of all of the living and nonliving things in a certain place.

What are the living and nonliving things in this photo? How do they interact?

Words to Know

When living and nonliving things *interact*, they affect each other. Sometimes, they cause changes in an *ecosystem*.

Think about a forest habitat. Birds live in the trees in a forest. The trees grow from soil. Worms live in the soil. Some birds eat the worms for food. All of these organisms in the forest interact. They are affected by nonliving things in the ecosystem, such as landforms and weather.

Adaptations in Living Things

Chapter 2

Think about the organisms in a pond habitat. A duck can glide across the water. A turtle can move from water to land. A fish can swim without having to come up for air. These animals are suited to a pond habitat. They have **adaptations** that help them get what they need there.

An adaptation is something that helps a living thing survive in its habitat. An organism's body parts are adaptations. They are inherited from parents. The shape of a duck's feet helps it paddle through water. A fish breathes through its gills.

Big Question
What are adaptations?

Vocabulary
adaptation, n. a body part or behavior that helps a living thing survive

Shiny duck feathers are coated in oil. The oil makes feathers waterproof. This is an adaptation. It helps a duck survive in a water habitat.

Adaptations Help Living Things Get Food

All living things need food to survive. Organisms have adaptations that help them get and eat food in their habitats.

A giraffe eats leaves on trees. However, it lives in a habitat that is mostly grass. There are only a few trees. A giraffe can use its long neck to reach the leaves in a tall tree. Its long neck is an inherited adaptation. This adaptation gives the giraffe an advantage. It can reach a food source that many other animals cannot reach.

A giraffe's long neck is an adaptation for survival.

A topi is another animal that lives in the same habitat. It is not tall like a giraffe. It eats grasses instead. Its snout is long and skinny. This adaptation helps it find grasses that are new and green. This trait gives the topi an advantage for surviving in the same habitat as a giraffe.

The topi's snout is an adaptation for survival.

Word to Know

An *advantage* is a condition that is helpful. A trait that is an advantage makes it easier for an organism to survive.

An animal that eats other animals often has to catch its food. These animals have different adaptations from animals that eat only plants.

An owl lives in a forest habitat. It hunts mice and other small animals at night. An owl has large eyes that help it see in the dark. It flies to the ground and catches food with its hooked claws, called talons. It uses its sharp, hooked beak to tear and eat its food. Large eyes, talons, and a hooked beak are adaptations. They help an owl catch and eat food in a dark forest habitat. They give the owl an advantage and help it survive.

The owl's wide wingspan is another adaptation. The owl can glide quietly to sneak up on animals that it catches for food.

Adaptations Help Living Things Defend Themselves

Organisms have adaptations that help them stay safe. A type of cactus called a saguaro lives in a hot, dry desert habitat. It has adaptations that save water. Its thick stem stores water. A tough, waxy outer layer keeps moisture inside the plant. Sharp spines make it hard for animals to eat the cactus. These adaptations give this cactus an advantage in a desert environment.

Spines make it difficult for an animal to take a bite out of a cactus.

A horned viper lives in a desert, too. Its colors help it blend in with the rocks and sand. This is called *camouflage*. Camouflage helps an animal stay safe in its habitat. Hawks and owls can't see the viper very well when they hunt.

The horned viper inherits the trait of its color pattern from its parents.

Think About Blending In

Camouflage looks different in different habitats. What kind of color patterns would be camouflage in different habitats? For example, what would a camouflaged animal look like on a sandy seafloor? What about in a leafy tropical tree?

The poison dart frog lives in a rain forest habitat. It is a very small animal and brightly colored. The poison dart frog is poisonous! Its bright colors are a warning. When other animals see it, they know not to eat the frog. The poison dart frog's warning colors are an adaptation. They give the frog a good defense against predators.

A poison dart frog's bright colors can also be a disadvantage. It cannot easily hide.

A capybara also lives in a rain forest habitat. A capybara is a furry animal about the size of a large dog. It has a different way of staying safe from predators. It jumps into nearby water! Most of its predators cannot swim. A capybara has webbed feet for swimming. It can hold

The capybara's webbed feet are an inherited trait.

its breath for five minutes. Both webbed feet and breath-holding are adaptations. They help the capybara protect itself from predators.

Adaptations Help Living Things Find Mates

Some animals have features that help them find mates. Male long-tailed widowbirds are colorful with bright feathers. They also have long tails to attract female widowbirds. Widowbirds that find mates can reproduce. They pass on their traits to young.

Female widowbirds are attracted to males with long tail feathers and colorful faces.

Giraffes' long necks are not only an advantage for reaching high leaves. Male giraffes use their long, strong necks as they fight each other for dominance. The winners in these contests are more successful at attracting female giraffes to mate with and reproduce.

Word to Know

Dominance is strength or power to succeed over others.

Living Things in Groups

Chapter 3

Some adaptations are physical traits. The ways organisms behave can be adaptations, too. A **behavioral trait** is a way an organism acts. Both kinds of adaptations help organisms survive in their habitats.

Some animals live in groups. Living in groups involves many behavioral traits. The animals interact with each other. Animals in a group work together in different ways. When they work together, more members of the group will survive in their habitat.

Big Question
How do organisms work together for survival?

Vocabulary
behavioral trait, n. a way an organism behaves, or acts

Wolves live in a family group called a pack. Individual and group behaviors help the pack survive. Behavioral traits are adaptations.

Hunting as a Group Helps Individuals Survive

Animals must have food to survive. It is easier for some kinds of animals to catch food, or prey, when they hunt together. Lions live in groups and hunt together. Such a group is called a pride.

Lions live in large, grassy habitats. Many of the animals they eat can run very fast. Lions do not run as fast as their prey. Chasing the prey does not work well for the lions. Instead, they watch it from far away. Then, the female lions form a circle around the prey. They quietly walk toward the prey until the circle gets smaller and smaller. Soon, the female lions pounce. The male lions join them once the prey has been caught. Hunting as a group helps each member survive. In most animal groups, males and females do different jobs.

Female lions move toward their prey. The male lion stays behind.

Dolphins are mammals that live in ocean habitats. They live in groups called pods. Sometimes a dolphin from a pod will hunt on its own. But dolphins often hunt together.

The fish that dolphins eat sometimes travel in large groups called schools. A pod of dolphins surrounds the school of fish. The dolphins herd the fish into a tight, thick ball. Then the dolphins take turns swimming through the ball of fish. They grab as many fish as they can in their mouths. The other dolphins in the pod keep the fish from escaping.

Dolphins work together to herd fish into a tight ball. How does this behavior make it easier for dolphins to get enough food?

Groups Provide Safety in Numbers

Sometimes, living in groups is safer than living alone. This is true in habitats where it is hard to hide. It is easy for a predator to attack one animal. It is harder to attack a whole group.

You learned that lions hunt in packs. Wildebeests are prey for lions. They live in the same wide-open grassland habitat. There are not many places to hide.

But wildebeests travel in very large herds. They move close together when a predator comes near. The older wildebeests stay on the outside of the group. The lions could be injured if they attack the adult wildebeests. They may decide to move on to easier prey.

Staying together in a herd is a behavioral trait. It is an adaptation that helps the animals survive in their habitat.

Meerkats live in southern African habitats that are hot and dry. Up to forty meerkats live in a group. A meerkat group is called a mob. All of the members of a mob have different jobs.

Some of the jobs help keep the mob safe. One meerkat keeps watch while others look for food. This meerkat is called a sentry. It stands on a mound of dirt or a rock and watches for predators. The sentry squeals if it senses danger. The other meerkats know to take shelter.

Other meerkats work together to dig holes and burrows. These burrows are under the ground. Here, adults and babies are safe from predators. They are also protected from the desert heat.

Can you tell which meerkat is the sentry?

Animal Groups Often Care for Young Together

Living in groups has other advantages. One of these is that animal parents have help caring for their young.

In a meerkat group, several helpers watch the young while others look for food. These helpers go all day without eating. Meerkats take turns babysitting so that the whole group has a chance to eat.

A group of elephants is called a herd. All of the females in a herd are from the same family. They all help care for the calves. This helps the younger females learn how to care for the babies they will have one day.

When there is danger, adult elephants surround the calves to keep them safe.

Nature and Changing Ecosystems

Chapter 4

Habitats are the places where organisms live. The living things in a place interact, which means they affect each other. The place itself also affects the living things. An **ecosystem** is all the living and nonliving things in a certain place and their interactions.

Sometimes an ecosystem changes. This change can happen fast. It can also happen very slowly. When a change happens, some living things may have trouble getting what they need. Some organisms will survive the change well. Some will have to work harder to meet their needs. Others will leave or even die.

Big Question

How do natural changes to an environment affect the organisms that live there?

Vocabulary

ecosystem, n. all the living and nonliving things in a place and their interactions

This bird gets shelter and finds food from trees in its ecosystem. What might happen to the bird if a change kills all the trees in the forest?

Natural Events Can Disrupt Ecosystems

Natural events cause changes in ecosystems. Events that cause damage are called natural hazards. These changes can happen suddenly.

Word to Know

A *natural hazard* is a natural event that causes negative changes to a habitat or ecosystem.

Hurricanes, tornadoes, earthquakes, fires, and floods are types of sudden natural hazards. Natural hazards can destroy trees and other plants. They can wash away sand, soil, and rocks. Natural hazards can force animals out of their nests, dens, and burrows. Many times, animals are forced to find a new place to live. These changes are natural events that can result in new ecosystems.

The plants and animals in a beach ecosystem are affected by a hurricane.

Living things can have a hard time finding food and water after a natural hazard. Plants that some animals eat may have washed away. Small prey may have disappeared. Sometimes animals are injured during a natural disaster. They may not be able to hunt. Water may be unsafe to drink.

The picture shows a forest habitat after a fire. Some living things have adaptations that help them survive fires. Many plants that burn may take a long time to grow back. Animals that live in trees in the forest have to move. Animals that eat the plants have to find other food.

It can take many years for the trees to grow back after a forest fire. But mosses, grasses, and other small plants may start growing on the bare ground quickly. How do these changes affect the animals that lived in the forest before the fire?

Invasive Species Can Disrupt Ecosystems

Sometimes organisms come to new habitats from other places. An **invasive species** is a type of organism that causes harm in a place where it does not normally live. Invasive species can cause change in habitats and ecosystems. These living things may grow and reproduce at a fast rate. Many do not have predators. They can be very hard to remove from a habitat.

Invasive species cause many problems. They take food and water from native plants and animals. Invasive insects may kill native trees and plants. Invasive plants can take up space that other plants need to grow. They can make it hard for sunlight to reach other plants, too.

Vocabulary

invasive species, n. an organism that causes harm in a place where it does not normally live

Word to Know

A *native* plant or animal is one that is found naturally in a habitat. Invasive species are not native to an area. They arrive in a habitat mainly because of human activity.

This stinkbug is an invasive insect. A large population of stinkbugs can destroy fruit and vegetable crops.

One invasive species is the lionfish. Lionfish are native to the southern Pacific Ocean and the Indian Ocean. Now they are found in the Atlantic Ocean, too. Lionfish have a big appetite. They eat more than forty different kinds of fish and other small ocean animals.

How does this affect an ocean ecosystem? Invasive lionfish eat so much that there is not enough food for native ocean animals. They also eat parrotfish, which are helpful to coral reef ecosystems. Parrotfish eat organisms that can harm coral reefs. When lionfish eat the parrotfish, the reef ecosystem may be harmed.

The lionfish is venomous. It does not have any natural predators where it has invaded. In such areas, the number of lionfish is growing quickly. It is harming ocean ecosystems.

Overpopulation Can Disrupt Ecosystems

Sometimes too many organisms in one ecosystem can cause the entire ecosystem to change. This is called overpopulation. When there are too many of a certain type of animal, the animals might consume all the food in their habitat. Then some of the population can starve.

A red tide is another example of overpopulation. In saltwater habitats, the algae sometimes reproduce too quickly. The algae turn the water red, orange, or pink. Oxygen can become scarce in the water. Fish can die when they eat the algae. Birds that eat the fish can become sick and die, too. The entire coastal ecosystem is affected.

> **Word to Know**
>
> *Overpopulation* happens when there are too many of one kind of species in an ecosystem.

Overpopulation of certain algae causes red tides. This can kill fish and other animals.

Humans Cause Ecosystem Change

Chapter 5

Natural changes can disrupt ecosystems. But not all changes are natural. People can cause changes to ecosystems, too. Like natural changes, human-caused changes can happen quickly or slowly. Cutting down trees for wood is a quick change. Human activities seem to be able to change Earth's climate. This is a slower change.

Big Question

How do people change environments in ways that affect the organisms that live there?

Human-caused changes can affect ecosystems in harmful ways. They can destroy animals' food and shelter. They can make soil unhealthy for plants. They can make water dirty and unsafe. When habitats change, living things must change, too. Some living things survive the changes. Other living things do not.

These trees were cut down in a forest habitat. How would this change affect animals that live in and around the trees?

Buildings and Roads Disrupt Ecosystems

Think about a building site near where you live. Humans need buildings for homes, hospitals, and offices. They need roads for transportation. Construction of roads and buildings can harm habitats.

Land must be cleared to make room for buildings and roads. Trees are cut down. Soil, rocks, and grass are removed. Plants that were food for animals disappear. Suddenly, animals are left with a smaller habitat. This can hurt animals that need space to hunt and to find shelter.

In many parts of the world, rain forests are being cleared to make room for buildings and crops. Orangutans live in some of these rain forests. They spend most of their lives in trees. They have nowhere to live or hide when the trees are removed. Many of them starve. Others are killed by predators.

Orangutans need rain forests for food and shelter. When humans cut down their habitat, orangutans cannot survive there.

Roads change habitats. Many animals **migrate** to find food and reproduce. When animals migrate, they move from one place to another in different seasons. Some species migrate very long distances, and others migrate short distances. Birds fly when they migrate. Other animals have to crawl, walk, or slither. Roads can form a barrier for these animals.

> **Vocabulary**
>
> **migrate, v.** to move from one place to another in different seasons

Roads can affect the population of a species. For example, some female turtles migrate from water habitats to lay eggs on land. The male turtles stay behind. Many female turtles are killed trying to cross roads. Over time, the number of female turtles decreases. There are not enough mates for the males. Fewer offspring are born.

Animals can be killed trying to cross roads.

Signs such as this one can remind drivers to watch for animals.

Human Use of Water Disrupts Ecosystems

Living things need water to survive. Humans are living things and need water, too. People use water for drinking, cooking, and washing. We use it for watering crops. We even use water to make electricity. But humans' use of water can affect ecosystems.

Humans build dams to collect water. A dam is a structure that blocks the flow of water in a river. Blocking rivers can cause problems for salmon.

Salmon are fish that migrate between two habitats. They hatch in rivers and move to the ocean. Then they come back to the rivers to reproduce. Dams can keep some salmon from migrating between rivers and oceans.

This is the Hoover Dam. It collects water from the Colorado River. People use the water for drinking, for farming, and to make electricity.

People living in cities and towns need fresh water. Communities often get fresh water from rivers and lakes. Sometimes people use so much that the water levels drop in the rivers and lakes.

Fish, turtles, frogs, and other animals live in freshwater habitats. When water levels drop, they have less space to live. The water can become cloudy and dirty. These changes harm the ecosystem.

Sometimes rivers and lakes dry up because of human water use. Plants and animals that live in the water die. Other animals that rely on these organisms for food are forced to move. They must find food and water somewhere else to survive.

Removing too much water from a river can cause it to dry up.

Changing Climate Disrupts Ecosystems

Climate is the pattern of weather over a long period of time. Many scientists have stated that human activities can cause changes in climate. These changes can cause problems for ecosystems.

Vocabulary

climate, n. the pattern of weather over a long period of time

Humans burn materials that release gases into the air. The gases become trapped in Earth's atmosphere (the layer of air that covers the planet). This causes Earth's temperature to rise. It causes ocean temperatures to rise, too. Climate change may make some areas colder than usual over time.

A changing climate affects sea ice in cold habitats. Polar bears live in these cold habitats. They stand on sea ice while they hunt for seals. As the climate warms, sea ice melts. It breaks into smaller pieces. These pieces float far out into the ocean. Polar bears cannot always reach them. Many have begun to starve.

A polar bear waits for a seal to come up through a hole in the sea ice.

Living Things Respond to Change

Chapter 6

Environments can change. Natural hazards cause disruptions. Humans cause changes, too. Living things must respond to changes to survive.

Remember that organisms have adaptations. Adaptations are traits that help living things survive in a certain habitat. Adaptations can be body parts. They can be behaviors, too.

When a habitat changes, an organism's adaptations may no longer be helpful. Organisms that stay in a changed habitat may struggle. They may no longer have shelter. They may have to go without food or water for a long time. Leaving may be the only way to survive.

Big Question

How do living things survive changes in their habitats?

Word to Know

Respond means to react to a change by doing something.

Can this red fox survive if more and more people move into its habitat?

Polar Bears Respond to Habitat Change

In Chapter 5, you learned about changes to a polar bear's habitat. Earth is getting warmer because of changes in climate. The amount of sea ice is decreasing in some areas. Polar bears need to stand on sea ice to catch seals. Now, some polar bears cannot find enough food.

Polar bears are large animals. They need a lot of food. Some of it is stored in their bodies as fat. The stored fat keeps the bears alive when food is harder to find. But the stored fat can only last so long. Polar bears rely mainly on seals for food. Polar bears eat very few other animals.

Some polar bears may have to walk a long way to find sea ice where they can hunt seals.

Polar bears are responding to less sea ice in different ways. Some polar bears swim or walk a long way to find sea ice. This uses up more energy than they can replace. Other polar bears don't eat during the warmest months. A few even try jumping into the ocean to catch a seal.

Polar bears must change their natural behavior. Some are able to survive. Others struggle. Some adult polar bears die because food is hard to get. Parents that cannot eat enough cannot feed their cubs either. The adults may lack the energy to find mates and reproduce.

Salmon Respond to Habitat Change

You also learned about salmon. These fish migrate between rivers and the ocean. When people dam rivers, salmon cannot easily migrate. Changes to Earth's climate also affect salmon.

Earth's warmer climate causes river temperatures to rise. Salmon need to live in cold water. Their bodies are not as healthy when the water is warm. This can harm their reproduction and growth. Some salmon have responded by changing when they migrate. They migrate earlier in the year to avoid warmer water temperatures.

Salmon lay their eggs in fresh water. Warmer river water can affect the survival of the eggs and young salmon.

Ecosystem Problems and Solutions

Chapter 7

You just read about changes in polar bear and salmon habitats. Human activities may have caused these problems. Many people want to find ways to fix these disruptions.

Conservationists are people who find solutions for problems in habitats and ecosystems. Conservationists work to save and protect plants and animals. They also work to protect air, water, and soil.

Conservationists want living things to have what they need to survive. On the next few pages, you will read about problems in certain habitats. Then you will read about solutions conservationists are using to try to fix them.

Big Question

What are some solutions to problems in habitats and ecosystems?

Vocabulary

conservationist, n. a person who works to protect plants, animals, habitats, and ecosystems

To conserve something means to protect it from harm. The act of protecting living things and their habitats is called conservation.

Hunters have killed so many white rhinos that there are very few left. The northern white rhino is just one species that conservationists are working to protect.

35

Using Beaver Dams to Repair Habitats

You know that human-built dams can cause problems in river habitats. They block salmon from migrating. They also cause changes in the water. These changes affect salmon populations. The salmon do not reproduce as often. Young salmon struggle to survive.

Conservationists found a natural way to help restore some salmon populations. They remove beavers from one habitat and put the beavers back into river habitats. This encourages the beavers to build natural dams of their own! The dams create ponds. The ponds change the habitat, and the salmon have more places to lay eggs. The natural beaver dams also result in water that is rich in nutrients. Young salmon grow faster in this water and are healthier.

Beavers have many positive effects on river habitats. Allowing beavers to build dams can help salmon lay eggs and survive.

Word to Know

Restore means to bring back something. When people restore a habitat or population, they repair damage that was done by nature or people.

Beavers help habitats and ecosystems in other ways, too. Their dams cause wetlands to form. A wetland is a place where the land is covered with water. Wetlands provide a habitat for a large number of living things. They can slow down forest fires and floods. They also filter water from rivers and streams to help keep it clean.

Beavers build dams with tree branches. This creates new habitats for some animals.

Conservationists are also putting beavers in river habitats high up in mountains. Their dams keep rivers and streams filled. The water flows down the mountains. It provides moisture for dry land at the bottom.

Building Safer Roads to Repair Habitats

Think about a large forest habitat. There is plenty of space for animals to live and hunt. Migrating animals can come and go. Now think of the same habitat with a road down the middle. The large habitat becomes two smaller habitats. There is not as much space for living things. Migrating animals must find a new path. Animals that try to cross the road may not survive.

People need roads. Living things must be able to roam and hunt safely. How can we conserve habitats and build roads? Conservationists came up with a solution. They designed wildlife crossings.

Animals walk over the bridge built by people. Fewer animals were killed after the bridge was built.

There are many kinds of wildlife crossings. Some go over roads. These can be used by larger animals, such as elk, deer, bears, and moose. Others are tunnels that go under roads. They can be used by smaller animals, such as tortoises, frogs, and badgers.

Some dams even have wildlife crossings called fish ladders. Fish ladders let migrating fish pass. People who design wildlife crossings make them as much like an animal's habitat as possible.

Wildlife crossings help animals migrate safely. They make roads safer for people, too. In many places with wildlife crossings, there are fewer accidents between cars and animals.

This is one example of a fish ladder. Fish swim up the ladder to get to the other side of the dam.

Solving Problems in Sea Turtle Habitats

Sea turtles spend most of their life in the ocean. They only come on land to lay their eggs in the sand.

Humans put sea turtle nesting sites in danger. People leave trash on the beach. Nesting turtles can get tangled in it. Building homes for people destroys the nesting habitat. Driving on beaches can crush the eggs. Wildlife steal the eggs for food.

Conservationists try many solutions to protect sea turtle eggs. Sometimes the nests are moved to quiet beaches. Special cages are built around the nests to warn drivers and keep out wildlife. Signs are put up so conservationists can find the nests and check on them. When the babies hatch, people make sure they have a clear path to the ocean.

Structures such as this one can protect a sea turtle's nest.

After they hatch, baby sea turtles make their way to the ocean.

Fossils and How They Form

Chapter 8

All living things in the past lived in some kind of habitat. They all lived in an ecosystem with other living and nonliving things. Even in the Age of Dinosaurs, living things depended on their habitats to live.

You probably know a few things about dinosaurs. You may know that they were reptiles. You may know that they were different shapes and sizes. How can people know about the habitats dinosaurs lived in?

Fossils provide evidence about dinosaurs and other living things from the past. A fossil is the remains of a living thing from long ago. Many fossils form in layers of rock. Fossils provide clues about what an organism ate. They can show where an organism lived and something about its habitat. They suggest an organism's size and shape. They can tell how it moved.

Big Question

What can fossils tell us about ancient habitats?

Vocabulary

fossil, n. the remains of a living thing from long ago, usually formed in layers of rock

Does this fossil look like any animals that are alive today?

Word to Know

Evidence is information that supports an argument. For example, sharp teeth in a dinosaur fossil support the claim that that dinosaur probably ate meat. Your own observations can be used as evidence. Facts presented by others can be used as evidence, too.

Bones Become Fossils

You may have seen pictures of entire dinosaur skeletons. When an animal dies, it becomes buried in mud. Its soft parts decompose, or break down. Hard parts, such as bones and teeth, are left behind as fossils. Scientists find these fossil parts. Then they put them together like a puzzle.

Word to Know

When part of an organism *decomposes*, it breaks down and becomes part of the environment.

How do bones form fossils? An animal's bones decompose very slowly. Water can seep into them. The water contains materials that fill spaces in the bones. Over time, the bones become more like rock. Like other fossils, they are buried in layers of mud. The mud hardens around the bones.

By studying these fossil bones, how might a scientist describe the habitat this dinosaur lived in?

Shells Become Fossils

Some of the most common kinds of fossils come from shells. Animals from long ago that had shells lived in or near water habitats. Many of them were clams or snails. The animals that lived inside the shells had soft bodies. When the animals died, the soft bodies decomposed. The hard shells remained.

Shell fossils can be whole shells. They form like fossil bones do. But most shell fossils are just an impression. An impression is like a picture of an organism. It is formed when an organism lies in mud for a long time. The organism breaks down, but its shape is left behind.

What if these fossil shells were found high on a mountain? What might this tell us about how ancient habitats have changed?

Plants Become Fossils

Animals are not the only organisms that form fossils. Plants form fossils, too. A plant can become buried in mud. More and more layers of mud form over it. The plant parts decompose. But the shape of the plant is left behind. The mud hardens into rock over time. A fossil is formed.

Scientists can learn a lot from plant fossils. These fossils are evidence of what habitats were there a long time ago. They can give clues about the habitat and climate. Scientists can also compare plant fossils with plants alive today to see how they have changed.

This is a fossil of a pine cone from a tree that lived long ago. Can such a fossil provide evidence of ancient habitats? What kinds of habitats do pine trees live in today?

Permafrost Can Preserve Soft Body Parts

As you have read, the soft parts of an organism decompose first. Then the hard parts are fossilized. But sometimes, an entire living thing becomes a fossil. This happens when it dies and is covered in mud that quickly freezes. This layer is covered with more layers of mud. The frozen layer is called permafrost. Permafrost preserves an organism's soft parts as well as hard parts. The soft parts include skin and muscle.

> **Word to Know**
>
> When something is *preserved*, it is kept in its original form.

The most common permafrost fossils are woolly mammoths. In one place, a frozen baby mammoth was found. The fossil formed about 40,000 years ago. Some permafrost fossils have been found underwater! This is evidence that the area was once dry land.

Permafrost fossils help scientists know what a woolly mammoth looked like when it was alive. If it looked like this, what kind of habitat did it live in?

Amber Can Preserve Organisms

Some living things from the past are preserved in amber. Amber forms from a sticky material that comes from trees. Insects and other living things became stuck in the sticky material. It hardened to form amber. The organisms were preserved inside it.

Living things preserved in amber look whole. You can see insects' body coverings, wings, legs, and antennae, for example. But the soft parts inside them are gone. Scorpions, ants, mosquitos, dinosaur feathers, plants, and even lizards are some of the organisms and body parts that have been preserved in amber.

If amber comes from trees, what does this tell us about the habitat of these trapped living things?

Fossil Clues About Changing Habitats

Chapter 9

Fossils give scientists clues about living things from the past. They also give scientists clues about past habitats. Fossil clues show that habitats in some places were different than they are today. Scientists can draw conclusions about past habitats by studying fossils of the organisms that lived there.

Scientists investigate fossils to understand how climate and landforms have changed over time. Fossils also show that there are many kinds of living things that are now **extinct**. This means that they no longer exist.

Big Question
What do fossils reveal about organisms, habitats, and change?

Vocabulary
extinct, adj. having no surviving members

This fossil is of a species of crocodile that lived in the past. How is it like this species that is alive today? How is it different?

47

Some Fossils Are Very Old, and Others Are More Recent

Some kinds of rocks form layers. The layers at the bottom are usually the oldest. The layers at the top are usually the youngest. Fossils are found in different rock layers. When scientists know the age of a layer, they can draw conclusions about the age of the fossils in the layer. They can compare older fossils to younger fossils. They can see how living things and their habitats have changed over time.

Some very old fossils are billions of years old. They were simple organisms with only a single cell. Other species, such as ancient birds, have been found in newer rock layers. Those layers are much younger—about 150 million years old.

Where is the oldest rock in these layers? Where is the youngest rock layer?

Evidence That Landmasses May Have Moved

The Arctic is a very cold place. The land and sea there are covered with snow and ice. But scientists found fossils of extinct alligators there. They know that alligators alive today only live in warm climates.

Coral reefs form in warm ocean water. But scientists have also found fossil coral in the Arctic. This is evidence that the rock on which the fossil coral was found was once in a warm ocean. Scientists suggest that the land on which these things once lived may have moved north into a cold climate.

Fossil coral provides a clue that land may have moved to a colder part of Earth's surface.

Evidence That Ancient Tropical Forest Habitats Have Changed

Colombia is a country in South America. Scientists studied fossils from a coal mine there. The current land is very rocky, and there are few trees. But the fossils show evidence that it was once a tropical forest.

Scientists found fossil ferns and palms. Ferns and palms are plants that live in warm, wet habitats today. Scientists compared the fossils with living plants to draw conclusions about the past habitat.

Coal itself is evidence of ancient plant life. Coal is a rock that forms from the remains of plants that lived in swamps and bogs. Swamps and bogs are common in tropical forests.

This is a fossil of a fern leaf. It could be from an area that was once a tropical forest.

Fossils Are Evidence That Life Has Changed over Time

Do you live in the western United States? There might be dinosaur bones in the rock under your backyard! Scientists know that giant dinosaurs roamed this land. Thousands of bones have been discovered there.

Dinosaurs left other clues besides bones. They left tracks in the mud. These tracks formed impression fossils. Scientists use the tracks to tell whether a dinosaur walked on two legs or four legs. Footprints also provide evidence about a dinosaur's size.

Scientists have also found dinosaur eggs and nests. These fossils provide clues that some kinds of dinosaurs cared for their young. Similarities between dinosaur eggs and bird eggs are evidence that birds alive today are related to dinosaurs.

These tracks show that dinosaurs roamed near this place in Colorado.

Fossils Are Evidence That Some Kinds of Living Things Are Now Extinct

Dinosaurs were reptiles that are now extinct. Giant mammals once roamed Earth, too. Mastodons, woolly mammoths, and giant ground sloths were giant mammals. These mammals are also extinct.

Some of these mammals lived in very cold places. Scientists know this because they found evidence of glaciers near the fossils. Glaciers are huge fields of ice and rock. These cold places suddenly began to turn warm. The warming temperature changed the habitats there. Many large mammals could not survive these changes. They became extinct.

> **Word to Know**
>
> A *glacier* is a huge field of ice and rock that slides very slowly downhill. Glaciers are described as rivers of ice.

Those giant mammal species are extinct today. But some of their remains are found in glacier ice. The remains include skin, fur, and bones. Scientists compare the fossil body parts to those of mammals alive today. They find evidence that living and extinct mammals are related.

This woolly mammoth most likely lived in a cold habitat.

Habitat Changes and Extinction

Chapter 10

Habitats are always changing. Some changes are small. Some are big. Some changes take a very long time. Some, such as floods and fires, happen fast. Changes in habitats affect living things. Living things may not be able to meet their needs in a habitat that has changed a lot. Many will have to move, and many may die.

Big Question

How do changes in habitats lead to the extinction of species?

If the destruction or change in a habitat is widespread and lasts a long time, an entire species may be in danger. A species becomes extinct when all of its members die. All the dinosaur, woolly mammoth, and mastodon species no longer roam the land because they are extinct. The habitats of these animals changed over a long time. All the individuals of these species were not adapted to the new habitats. They could not respond in ways that kept them safe.

Dinosaur habitats changed over time. As a result, dinosaur species became extinct.

53

Habitat Change Caused Dinosaur Extinction

How did dinosaurs become extinct? Scientists have two main ideas. The evidence comes from rocks and fossils.

One idea is that a large asteroid from space crashed into Earth. The crash caused dust and smoke to rise into the air and block the sun. The world became cold and dark. Plants could not grow because they did not have sunlight to make food. Plant-eating dinosaurs died. Then meat-eating dinosaurs died, too.

The other idea is that a large volcano erupted. Smoke and ash blocked the sun. Acid rain fell from the sky. Dinosaurs died from lack of food.

These are only two ideas, and something else may have led to the extinction of dinosaurs. No matter what the cause, when a habitat or environment changes, some living things may not be able to survive and reproduce.

When a volcano erupts, smoke and ash fill the air and could result in habitat change.

Habitat Change Can Cause Widespread Extinction

When the dinosaurs got wiped out, so did many other species. This type of event is called a mass extinction because it causes the deaths of many organisms at once. Other mass extinctions have happened on Earth. Fossils give scientists clues about these events.

Natural events and changes have caused mass extinctions in the past. Some scientists have suggested that we are now in a time of a new mass extinction. This one, they suggest, is caused by human activity.

An asteroid like this one may have crashed into Earth and caused a mass extinction.

Changes in Habitat Have Caused Extinction Recently

In 1900, there were one million western black rhinos on Earth. In 2001, there were only about 2,300. A few years later, black rhinos were extinct. Hunters killed western black rhinos for their horns. Their habitats were destroyed for building and farming. The western black rhino could not survive these changes.

The Steller's sea cow became extinct about 250 years ago. It ate a plant-like organism called kelp. Sea otters lived in these kelp forests. They ate sea animals called urchins that eat kelp, too. The otters in that area were hunted for their fur, and they disappeared. The urchin population grew because the urchins had no predators. The urchins quickly ate all the kelp. Sea cows died because they could not find food. Again, habitat change can lead to the death of all the individuals of an entire species.

This black rhino is a relative of the western black rhino. It might become extinct, too.

The passenger pigeon was a North American bird. People thought its meat was tasty. They also thought these birds were pests. Hunters shot the birds and poisoned them. They destroyed the birds' nests. Passenger pigeons became extinct in the early 1900s.

Plants also can become extinct because of habitat change. The Saint Helena olive tree is one. It lived on an island in the Atlantic Ocean. People cut down these trees for wood. The few trees that were left had trouble reproducing. Scientists tried to save the trees. They grew young trees in a lab. But these trees could not survive in the wild. The last Saint Helena olive tree died in 2003.

Human activity caused the extinction of the passenger pigeon.

Human activity caused the extinction of the Saint Helena olive tree.

Habitat Change Threatens Many Living Things

Some species are close to becoming extinct. They are endangered. An **endangered species** is one that has almost disappeared. Conservationists try to help endangered species survive. But some of them are likely to die anyway.

> **Vocabulary**
>
> **endangered species,** n. a species that is at risk of becoming extinct within a few years

The Sumatran elephant is an endangered species. It lives in a forest habitat. Its habitat is being cleared to make room for farms and buildings. The elephants are also killed for their tusks. There are only about 2,500 Sumatran elephants left in the wild.

The Amur leopard lives in forests in Asia. There are fewer than one hundred Amur leopards left in the wild. People kill these leopards for their fur. People also hunt their prey for food. Construction and forest fires have changed the leopards' habitat, too.

The Amur leopard is in danger of becoming extinct.

Glossary

A

adaptation, n. a body part or behavior that helps a living thing survive (7)

B

behavioral trait, n. a way an organism behaves, or acts (13)

C

climate, n. the pattern of weather over a long period of time (30)

conservationist, n. a person who works to protect plants, animals, habitats, and ecosystems (35)

E

ecosystem, n. all the living and nonliving things in a place and their interactions (19)

endangered species, n. a species that is at risk of becoming extinct within a few years (58)

extinct, adj. having no surviving members (47)

F

fossil, n. the remains of a living thing from long ago, usually formed in layers of rock (41)

H

habitat, n. the natural place where an organism lives (1)

I

invasive species, n. an organism that causes harm in a place where it does not normally live (22)

M

migrate, v. to move from one place to another in different seasons (27)

S

survive, v. to stay alive (1)

CKSci™
Core Knowledge SCIENCE™

Editorial Directors
Daniel H. Franck and Richard B. Talbot

Subject Matter Expert

Christine May, PhD
Associate Professor of Biology
James Madison University
Harrisonburg, Virginia

Illustrations and Photo Credits

Adrian Wojcik / Alamy Stock Photo: 37
age fotostock / SuperStock: 20
Alex Witt / Alamy Stock Photo: 31
ardea.com / Mary Evans / Colin / Pantheon / SuperStock: 15
Aurora Photos / SuperStock: 4a
BlueGreen Pictures / SuperStock: Cover A, 40b
Ciobaniuc Adrian Eugen / Alamy Stock Photo: 6
Clem HaagnerPant / Pantheon / SuperStock: i, iii, 17
Corbin17 / Alamy Stock Photo: 52
Cultura Limited / Cultura Limited / SuperStock: 2b
DAVID HERRAEZ / Alamy Stock Photo: Cover B, 42
Delta Images / Image Source / SuperStock: 16
Design Pics Inc / Alamy Stock Photo: 2a
dotted zebra / Alamy Stock Photo: 45
Eric Sohn Joo Tan/ BIA / Minden Pictures / SuperStock: 12a
F1 ONLINE / SuperStock: 8a
Francois Gohier / Pantheon / SuperStock: 51
fStop Images / SuperStock: Cover C, 3a
Geoffrey Peter Kidd / age fotostock / SuperStock: 50
George Ostertag / Alamy Stock Photo: 44
Hans Blosseyo / imageBROKER / SuperStock: 38
Hans-Peter Merten / Mauritius / SuperStock: 10a
imageBROKER / Alamy Stock Photo: 13, 35
imageBROKER / SuperStock: 4b, 7, 9, 27a, 39
Ion Alcoba Beitia / PhotoMedia / Don Paulson Photography / SuperStock: 24
Jacques Jacobsz / Alamy Stock Photo: 18
Jim M Macdonald - Commercial / Alamy Stock Photo: 40a
Juan Carlos Muñoz / age fotostock / SuperStock: 48
Juniors / SuperStock: 56

Kevin Schafer / Minden Pictures / SuperStock: 3b
Konstantin Kirillov / Alamy Stock Photo: 41
Mattteo Festi / Alamy Stock Photo: 29
Minden Pictures / SuperStock: 34
National Geographic Image Collection / Alamy Stock Photo: 10b, 26
Nature Picture Library / Alamy Stock Photo: 12b
NHPA / SuperStock: 57a
Nikreates / Alamy Stock Photo: 49
Oleg Lopatkin / Alamy Stock Photo: 47b
Panther Media GmbH / Alamy Stock Photo: 5b, 11a, 22
Phil Robinson / age fotostock / SuperStock: 46
Pixtal / SuperStock: 1
PRILL Mediendesign / Alamy Stock Photo: 43
Radius / SuperStock: 5a, 32–33
Robert McGouey / age fotostock / SuperStock: 27b
RooM the Agency / Alamy Stock Photo: 14
Science Photo Library / SuperStock: 25, 55
SeaTops/imageBROKER / imageBROKER / SuperStock: 23
Stocktrek Images / Stocktrek Images / SuperStock: 53
Suzi Eszterhas / Minden Pictures / SuperStock: 8b
Tammy Wolfe / age fotostock / SuperStock: 36
The History Collection / Alamy Stock Photo: 57b
Thomas Marent / Minden Pictures / SuperStock: 11b
Tom Brakefield / Purestock / SuperStock: 58
Universal Images / SuperStock: 47a
Vladimir Seliverstov / Alamy Stock Photo: Cover D, 30
Werner Layer / Mauritius / SuperStock: 19
Westend61 / SuperStock: 21, 28, 54